Fruits

and

FARTS

Written & illustrated
by Alicia Kirk

ISBN: 978-0-9756461-0-6

Published by Alicia Kirk
Australia
First edition 2024
Written and illustrated by Alicia Kirk

for my family & friends
with love

apricot

banana

grapes

watermelon

Do you have a favourite **Fruit**?

What does it **Smell** like?

apple

pineapple

cherry

lime

lemon

kiwifruit

peach

orange

plum

nectarine

pear

strawberry

tomato

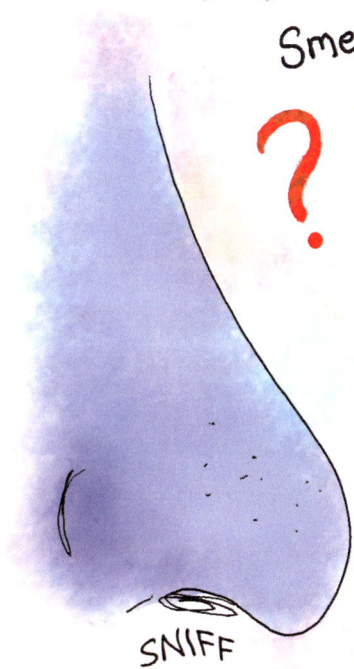

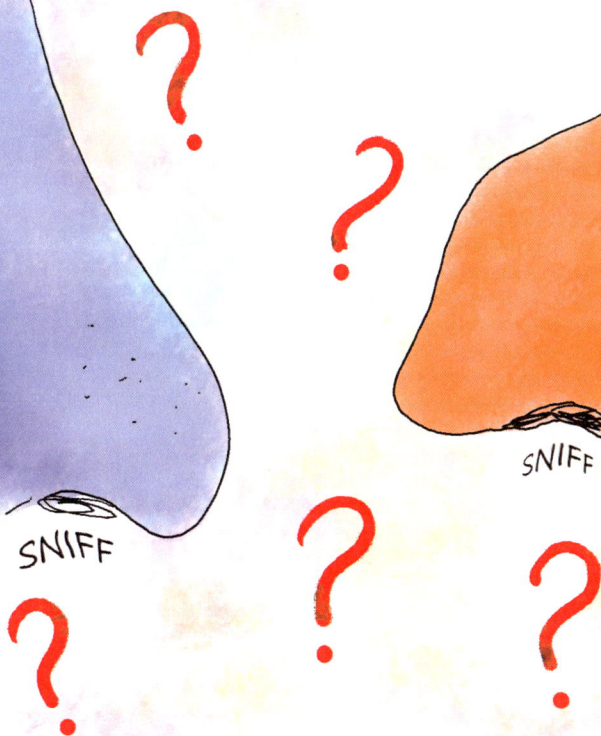

There are molecules in the fruit....

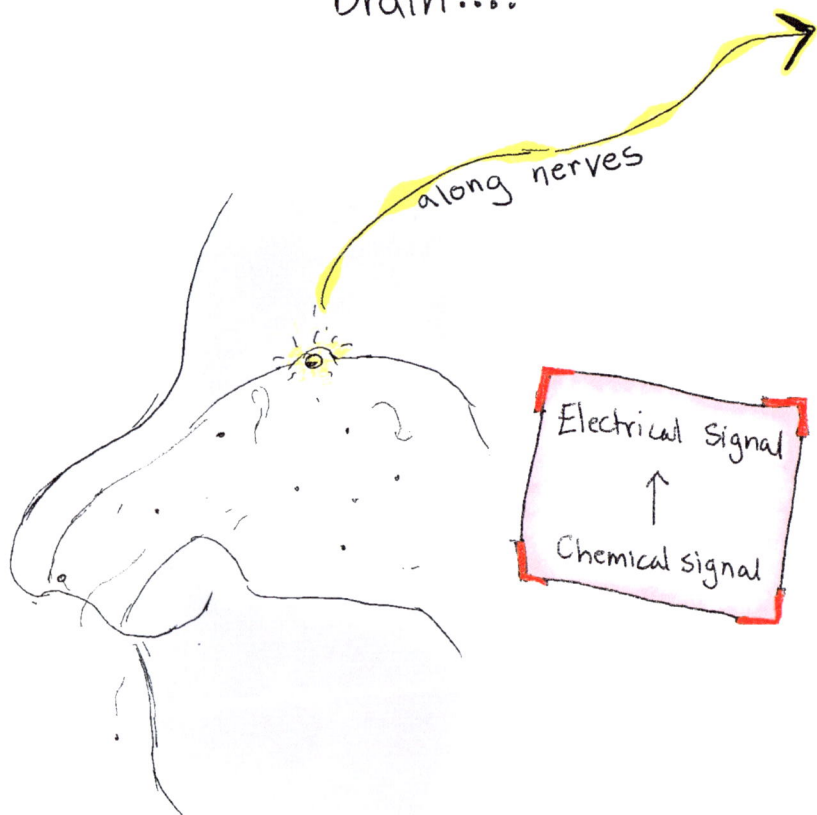

They Send
Signals to the
brain....

along nerves

Electrical Signal
↑
Chemical signal

...where your
brain processes
the smell.

But what are these molecules?

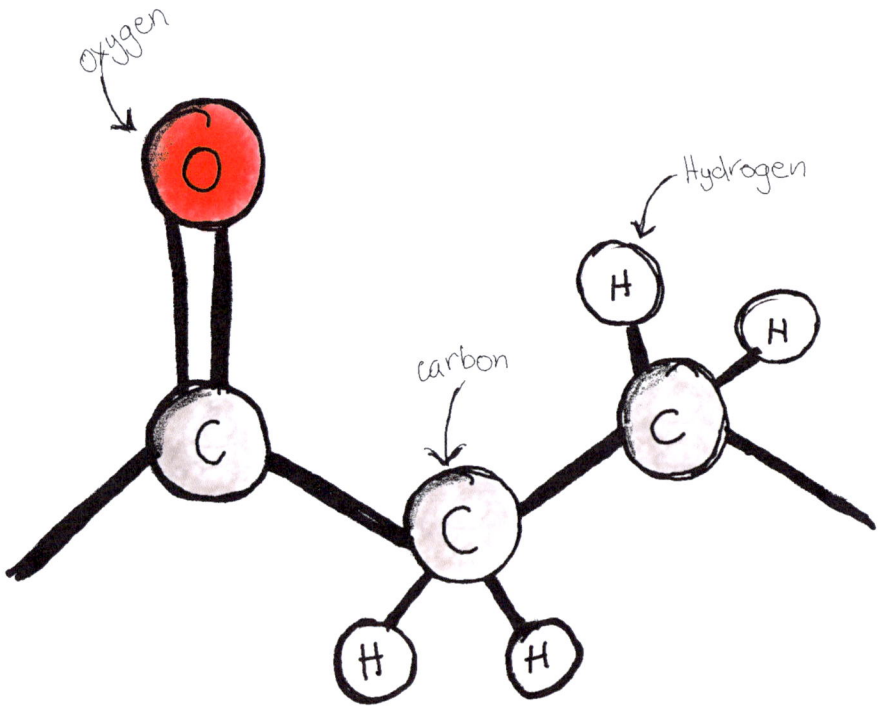

oxygen

O

Hydrogen

H

H

carbon

C

C

H

H

They are made up of atoms!

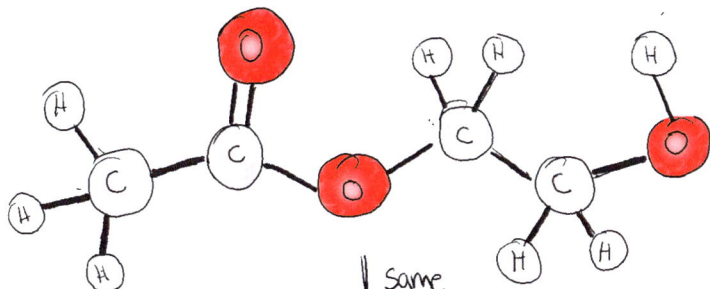

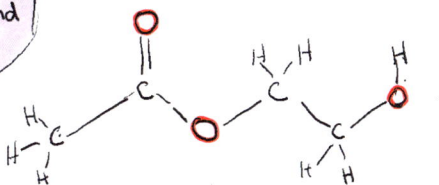

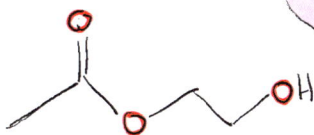

The main smelly molecules are esters!

They have this pattern:

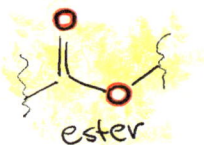

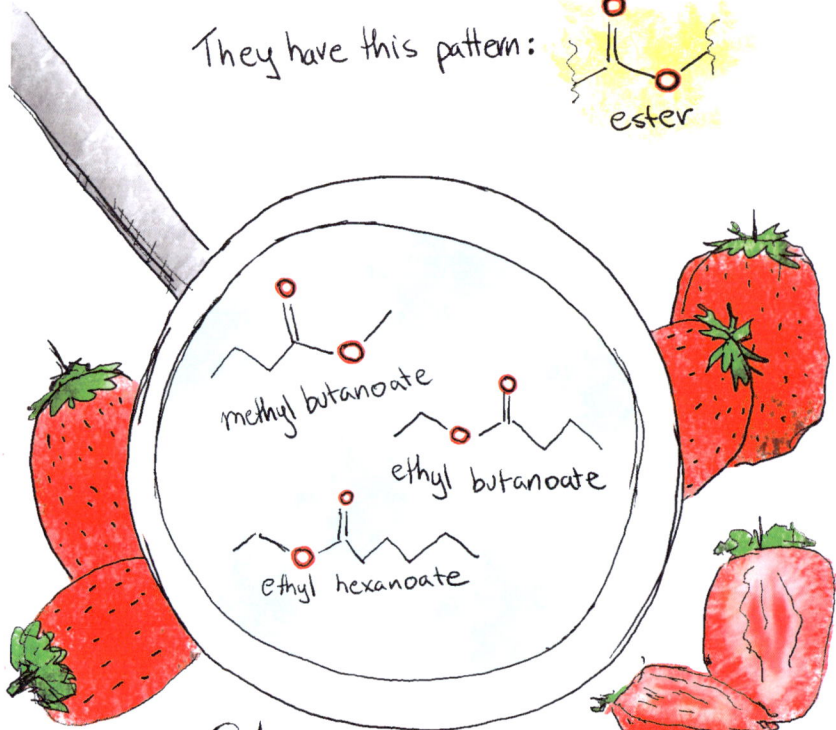

ester

methyl butanoate

ethyl butanoate

ethyl hexanoate

Strawberries

There are 100s of different molecules that make up the fruity smells we know!

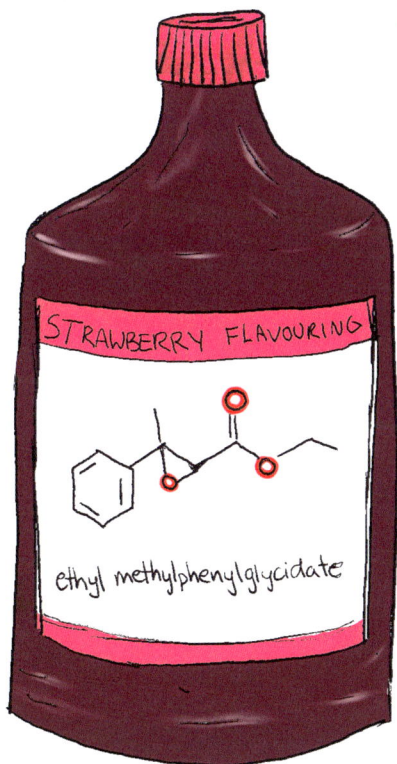

STRAWBERRY FLAVOURING

ethyl methylphenylglycidate

Did you know ...the ester used for **Strawberry flavour** isn't in Strawberries!

Here are
some more

Esters

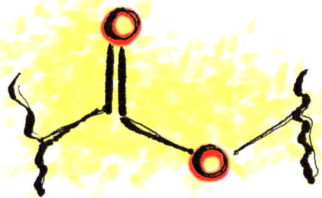

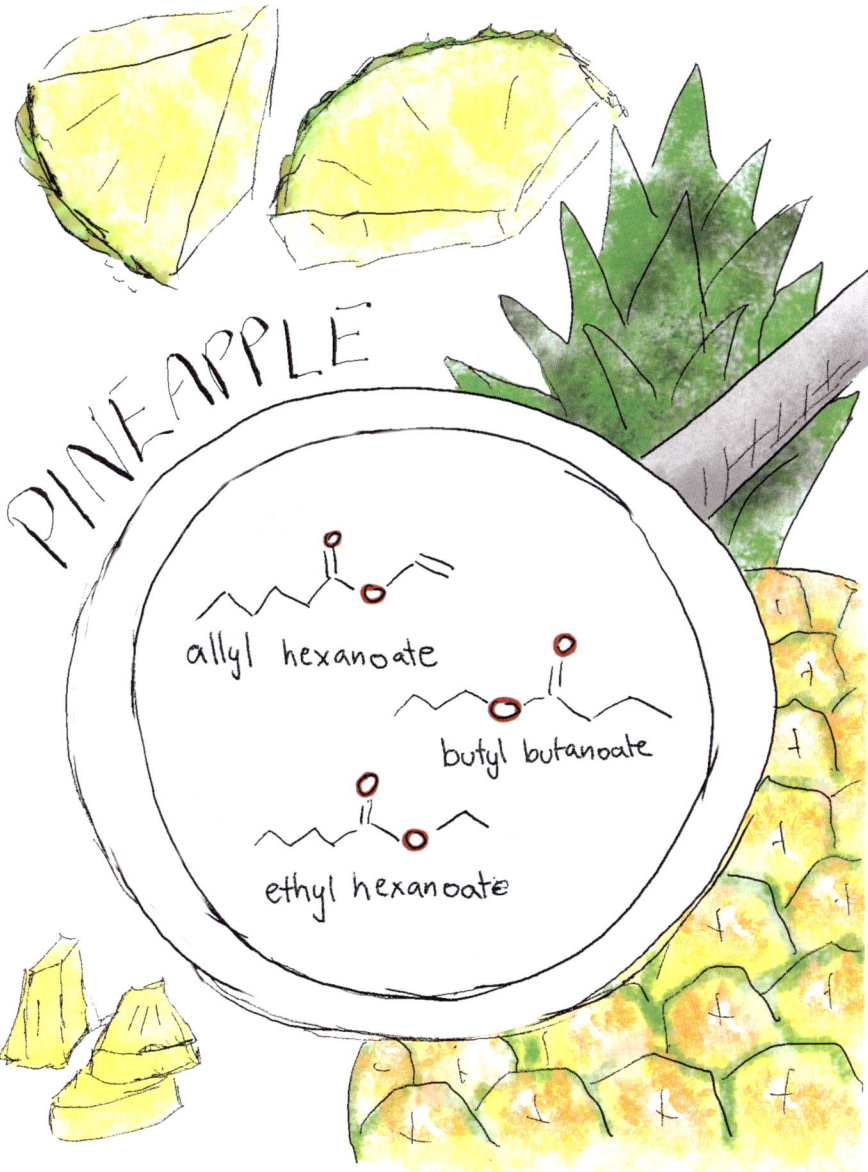

PINEAPPLE

allyl hexanoate

butyl butanoate

ethyl hexanoate

the furry one

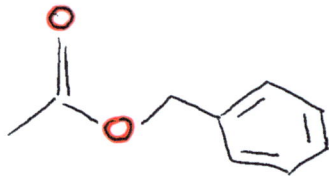

benzyl acetate

PEACH

BANANA!

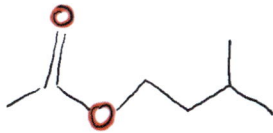

isopentyl acetate
"banana oil"

Apple

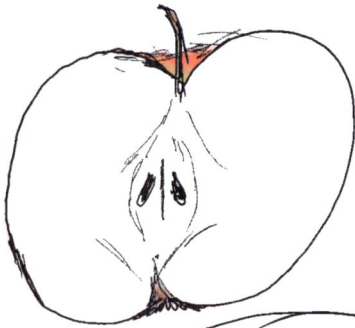

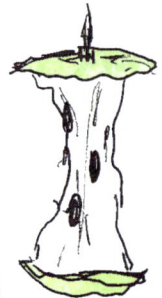

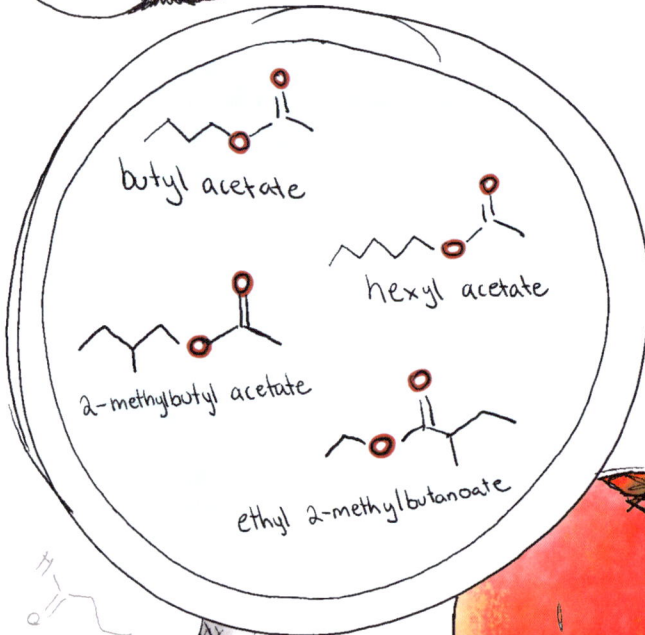

butyl acetate

hexyl acetate

2-methylbutyl acetate

ethyl 2-methylbutanoate

hexanal

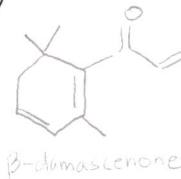

β-damascenone

octyl acetate

ORANGE

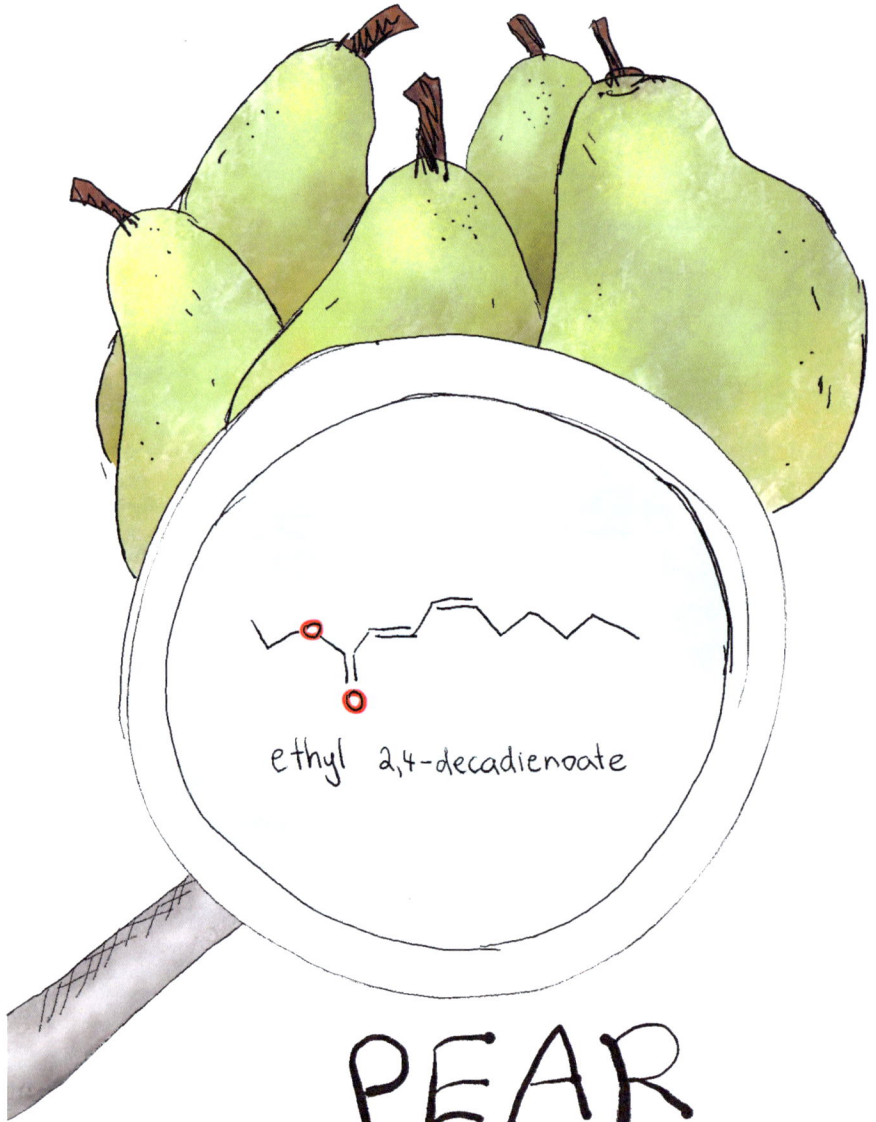

ethyl 2,4-decadienoate

PEAR

Raspberry

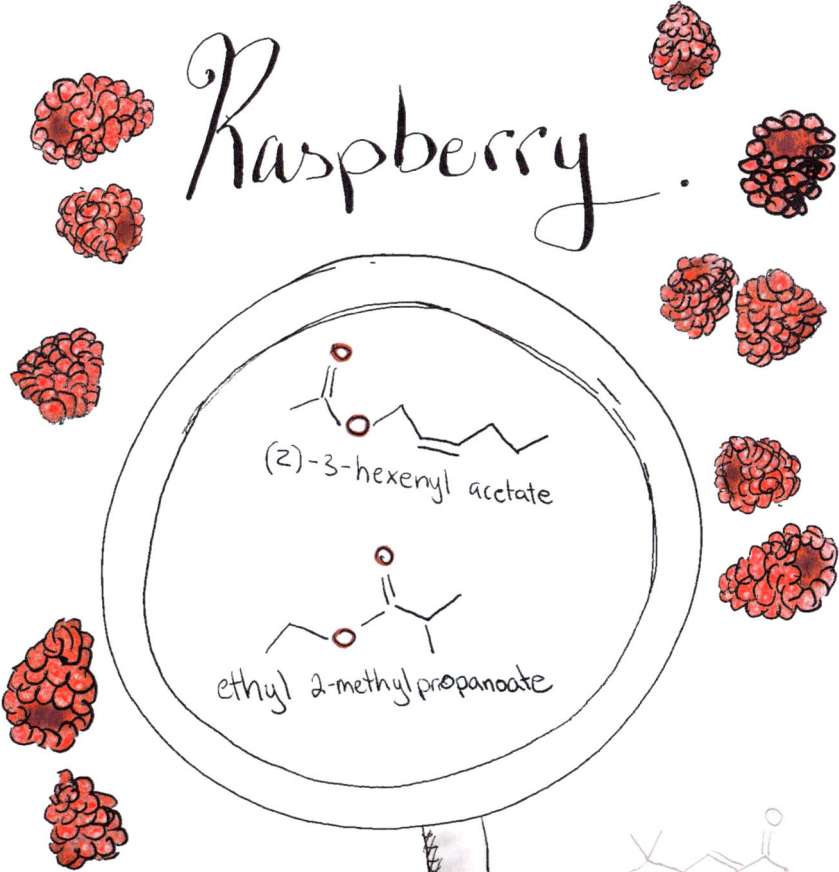

(2)-3-hexenyl acetate

ethyl 2-methylpropanoate

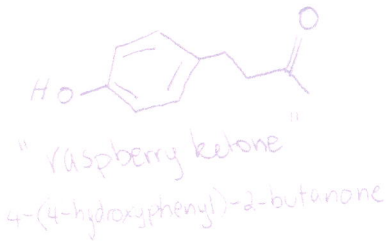

"raspberry ketone"
4-(4-hydroxyphenyl)-2-butanone

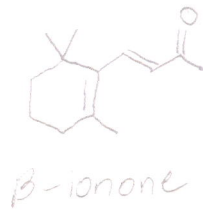

β-ionone

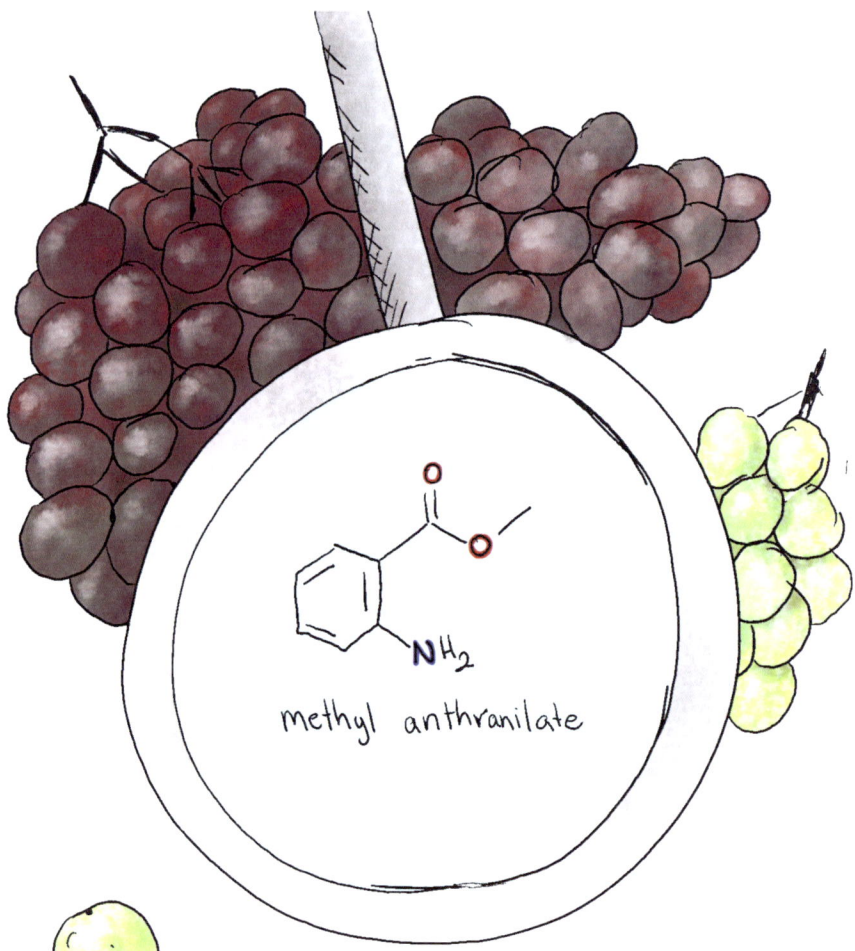

methyl anthranilate

Grapes

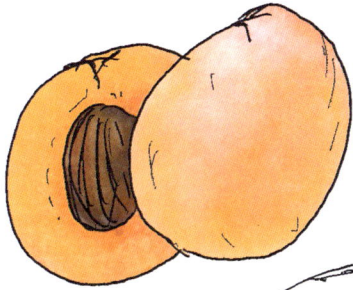

Cyclic esters are called **Lactones**

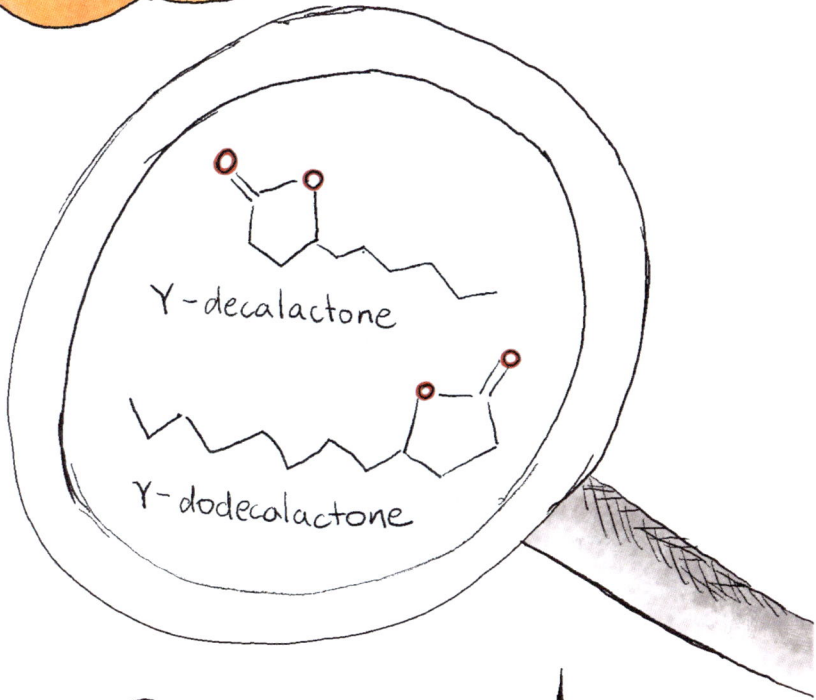

γ-decalactone

γ-dodecalactone

Apricot

....But not everything

smells fruity....

Have you ever smelt a **FART**?

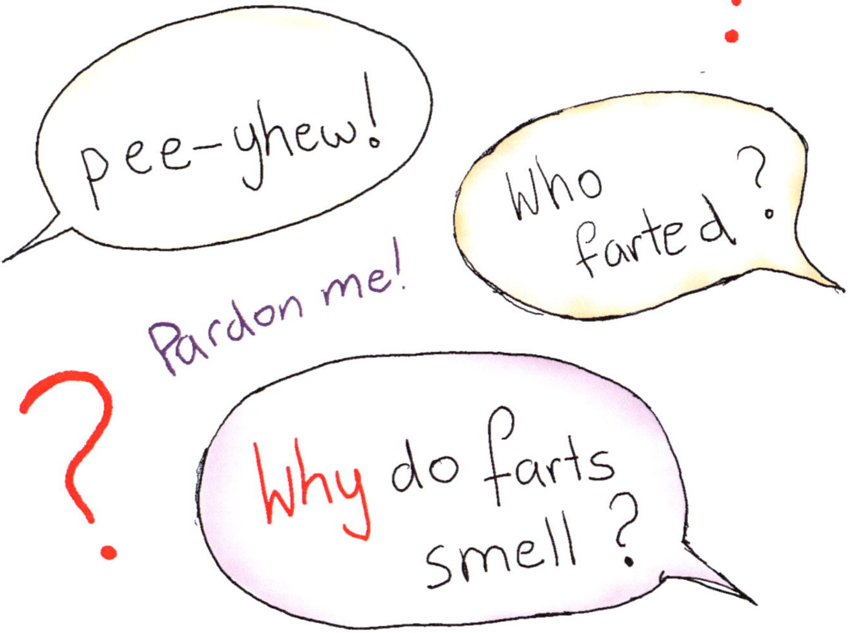

Well ... there are molecules in farts too....

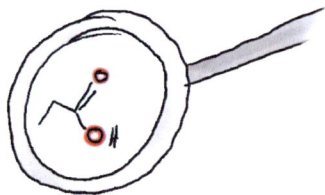

... which your nose can detect.....

... that sends signals to your brain....

... which might make you object....

... to a farty perfume!

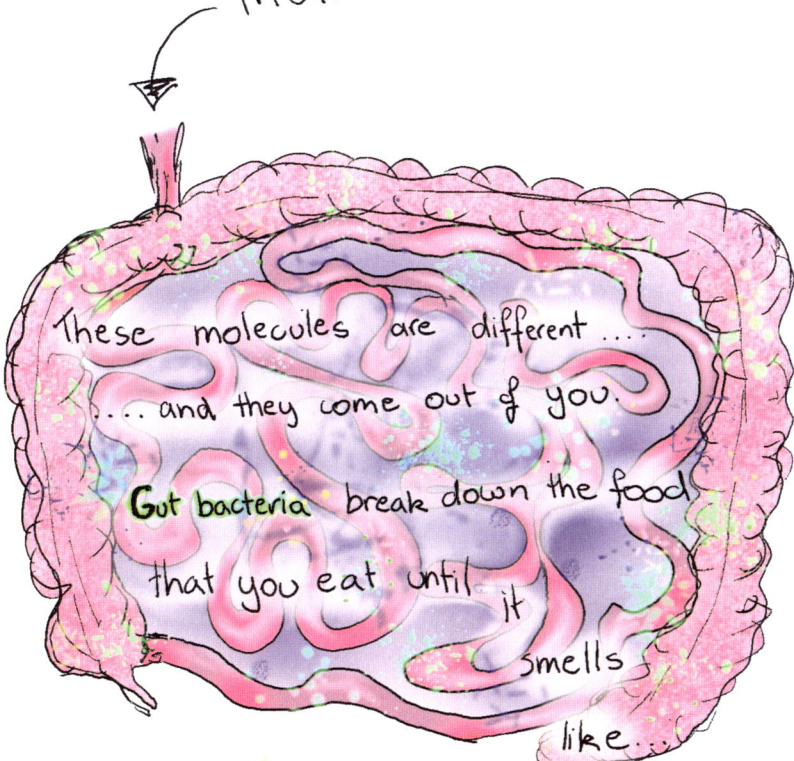

FOOD
molecules

These molecules are different ....
.... and they come out of you.
Gut bacteria break down the food
that you eat until it
smells
like...

INTESTINES
Large + small

POO!

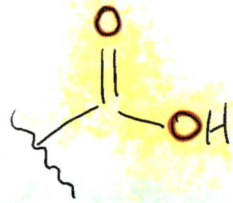

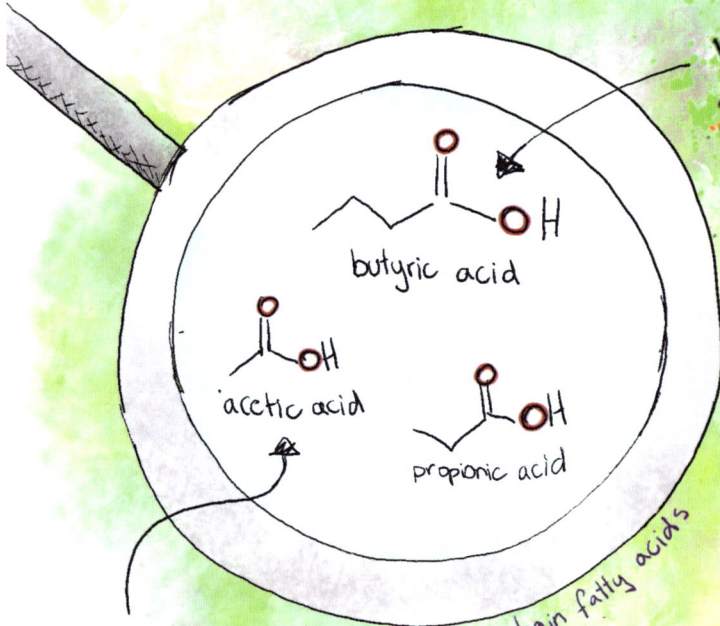

16
S
32.06

Stinky Sulfur compounds!

like rotten 〟 eggs

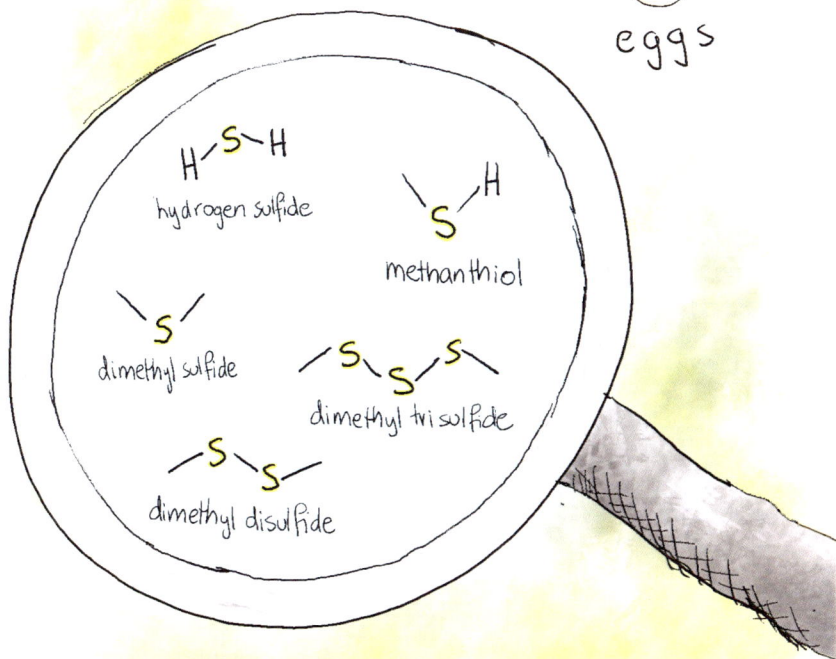

H–S–H
hydrogen sulfide

S–H
methanthiol

S
dimethyl sulfide

S–S–S
dimethyl trisulfide

S–S
dimethyl disulfide

... the sulfur comes from proteins that you eat
and vegetables like  and

onion   garlic...

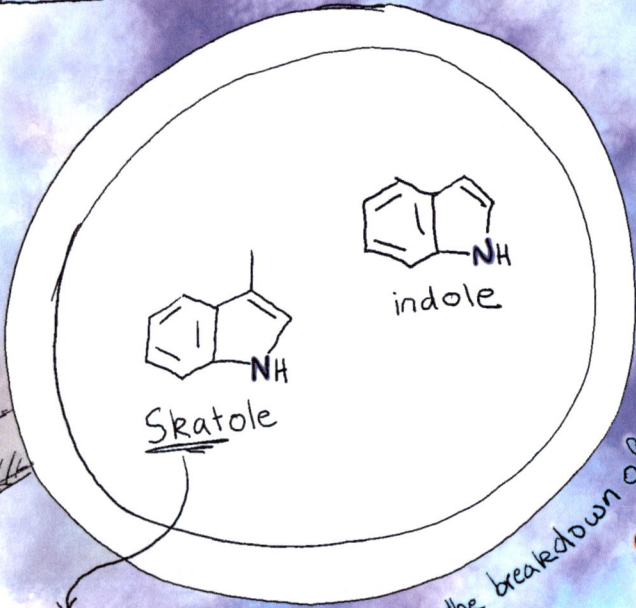

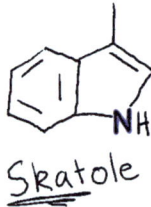

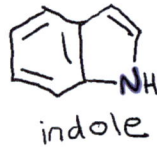

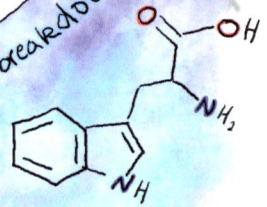

phenolic compounds
have this pattern:

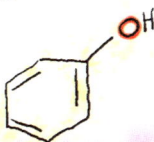

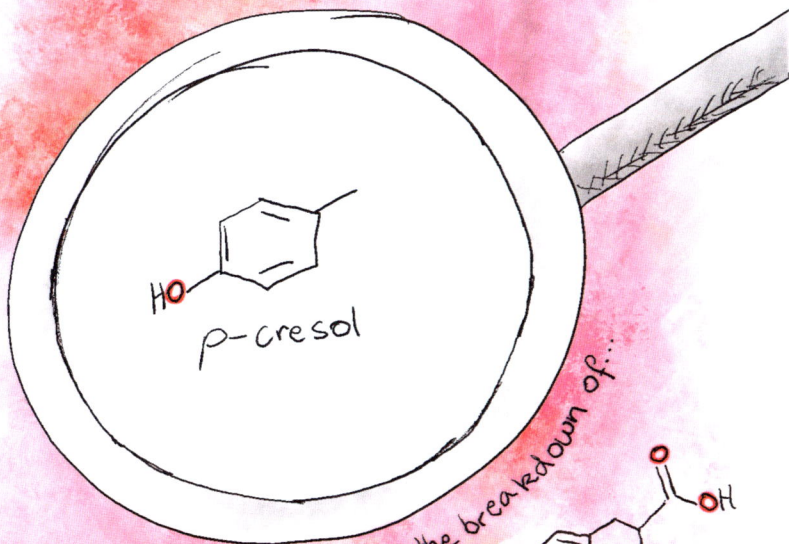

p-cresol

... from the breakdown of...

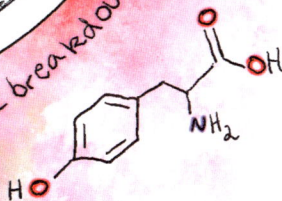

tyrosine

(another amino acid)

p-cresol
attracts
female mosquitoes

But did you know most of the gas in your farts doesn't smell at all!?

Some of it comes from air you've swallowed!

GULP
GULP
GULP

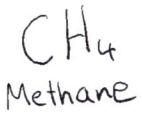

Fruity farts

or

FARTY fruits....

Have you heard of **DURIAN**?

It's a foul smelling fruit!

Fruity **ester**

It's banned from public spaces in some asian countries!

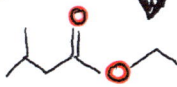

ethyl 2-methylbutanoate

1-(ethylsulfanyl)ethanthiol

AND

Stinky **Sulfur** compound

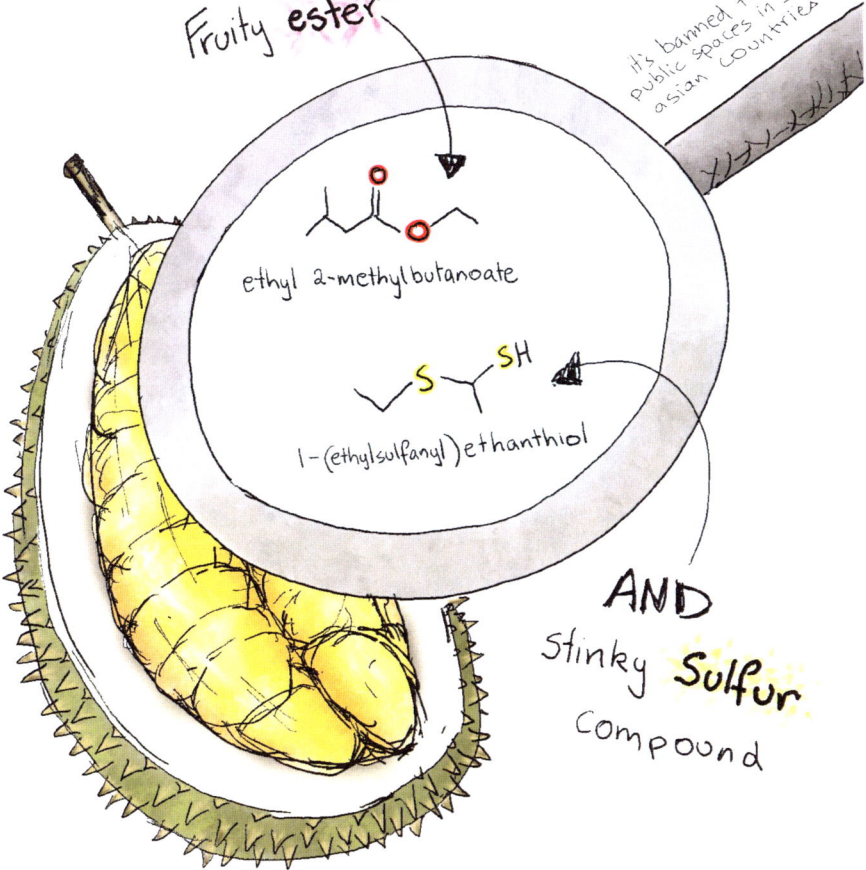

There are lots
of things you'll
smell
when you are

sniff-sniffing

around.

Now you might know
the *name*
of the **compound!**

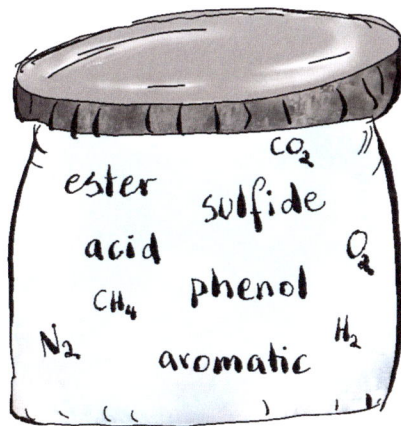

What molecules would you add to your perfume?

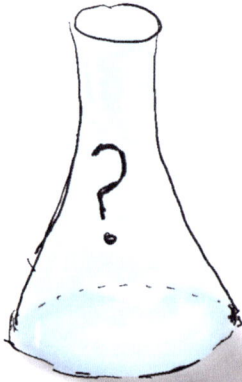

Design your own concoction:

Filled with esters or acids or sulfides or anything your imagination can think up!

MORE SMELLY MOLECULES:

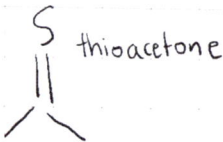

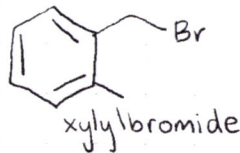

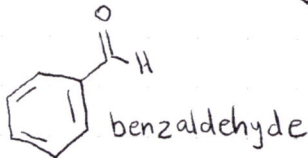

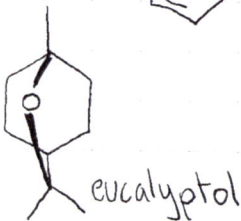

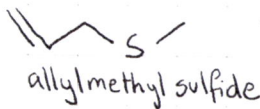

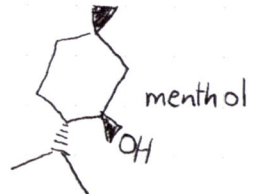

thioacetone

xylylbromide

benzaldehyde

menthol

OH

eucalyptol

allylmethyl sulfide